2nd EDITION

Activity Book
Foundation C

Make 5

Date: _____

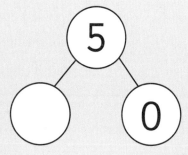

5
3 ◯

☐ + ☐ = 5

☐ + ☐ = 5

5
1 ◯

☐ + ☐ = 5

☐ + ☐ = 5

5
◯ 0

☐ + ☐ = 5

☐ + ☐ = 5

2

Subtract from 5

Date: _____

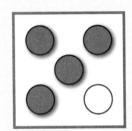

$$5 - 1 = \boxed{}$$

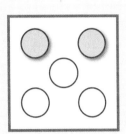

$$5 - 3 = \boxed{}$$

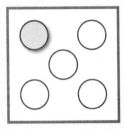

$$5 - 4 = \boxed{}$$

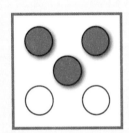

$$5 - 2 = \boxed{}$$

$$5 - 0 = \boxed{}$$

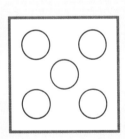

$$5 - 5 = \boxed{}$$

Teacher's notes

For each array, complete the subtraction number sentence to show how many counters are left.

Date: _____

Make 6

6
5   ◯

☐ + ☐ = 6

☐ + ☐ = 6

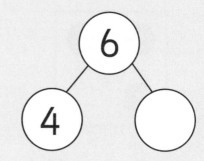

6
4   ◯

☐ + ☐ = 6

☐ + ☐ = 6

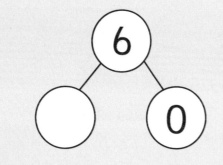

6
◯   0

☐ + ☐ = 6

☐ + ☐ = 6

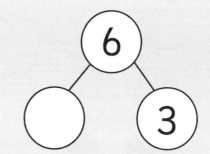

6
◯   3

☐ + ☐ = 6

Complete each part-part-whole model: write the missing 'part'. Then write the addition number sentences to match.

Date: _____

Subtract from 6

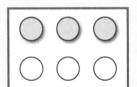

6 – 2 = ⬜

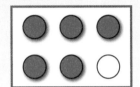

6 – 1 = ⬜

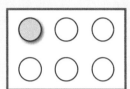

6 – 5 = ⬜

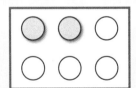

6 – 3 = ⬜

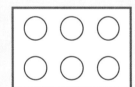

6 – 6 = ⬜

6 – 4 = ⬜

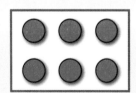

6 – 0 = ⬜

Date: _____

Make 7

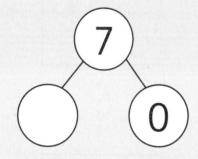

$$\boxed{} + \boxed{} = \boxed{7}$$

$$\boxed{} + \boxed{} = \boxed{7}$$

$$\boxed{} + \boxed{} = \boxed{7}$$

$$\boxed{} + \boxed{} = \boxed{7}$$

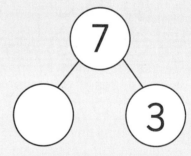

$$\boxed{} + \boxed{} = \boxed{7}$$

$$\boxed{} + \boxed{} = \boxed{7}$$

$$\boxed{} + \boxed{} = \boxed{7}$$

$$\boxed{} + \boxed{} = \boxed{7}$$

Teacher's notes

Complete each part-part-whole model: write the missing 'part'. Then write two addition number sentences to match.

Date: _____

Subtract from 7

7 – 5 =

7 – 0 =

7 – 3 =

7 – 4 =

7 – 1 =

7 – 2 =

7 – 7 =

7 – 6 =

Teacher's notes

For each array, complete the subtraction number sentence to show how many counters are left.

7

Make 8

8
5 ◯

◻ + ◻ = 8

◻ + ◻ = 8

8
◯ 0

◻ + ◻ = 8

◻ + ◻ = 8

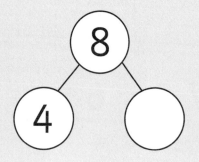

◻ + ◻ = 8

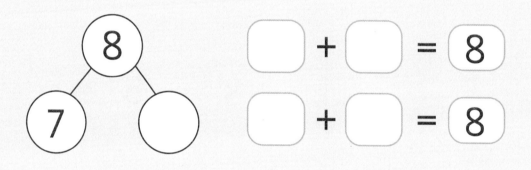

Teacher's notes

Complete each part-part-whole model: write the missing 'part'. Then write the addition number sentences to match.

Subtract from 8

$$8 - 7 = \boxed{}$$

$$8 - 2 = \boxed{}$$

$$8 - 5 = \boxed{}$$

$$8 - 0 = \boxed{}$$

$$8 - 1 = \boxed{}$$

$8 - 8 = \boxed{}$

$8 - 6 = \boxed{}$

$8 - 3 = \boxed{}$

$8 - 4 = \boxed{}$

Teacher's notes

For each array, complete the subtraction number sentence to show how many counters are left.

Date: _____

Make 9

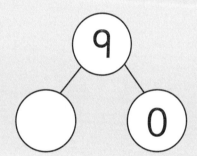

```
□ + □ = 9
□ + □ = 9
```

```
□ + □ = 9
□ + □ = 9
```

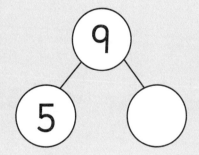

```
□ + □ = 9
□ + □ = 9
```

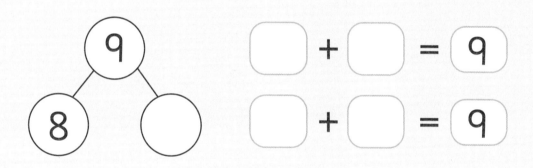

Teacher's notes

Complete each part-part-whole model: write the missing 'part'. Then write two addition number sentences to match.

13

Subtract from 9

$$9 - 5 = \boxed{}$$

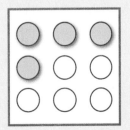
$$9 - 0 = \boxed{}$$

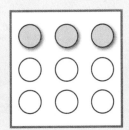
$$9 - 6 = \boxed{}$$

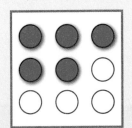
$$9 - 4 = \boxed{}$$

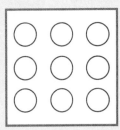
$$9 - 9 = \boxed{}$$

14

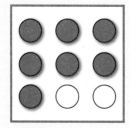

$9 - 2 = \boxed{}$

$9 - 8 = \boxed{}$

$9 - 3 = \boxed{}$

$9 - 7 = \boxed{}$

$9 - 1 = \boxed{}$

Teacher's notes

For each array, complete the subtraction number sentence to show how many counters are left.

Date: ____________________

Make 10

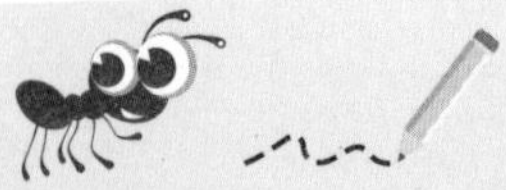

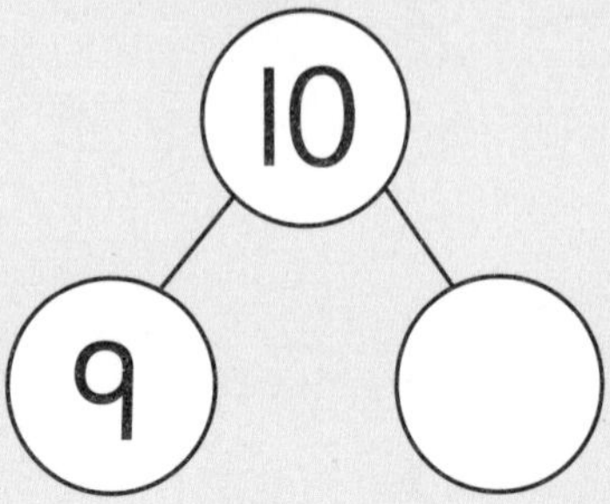

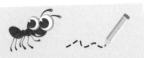

☐ + ☐ = 10

☐ + ☐ = 10

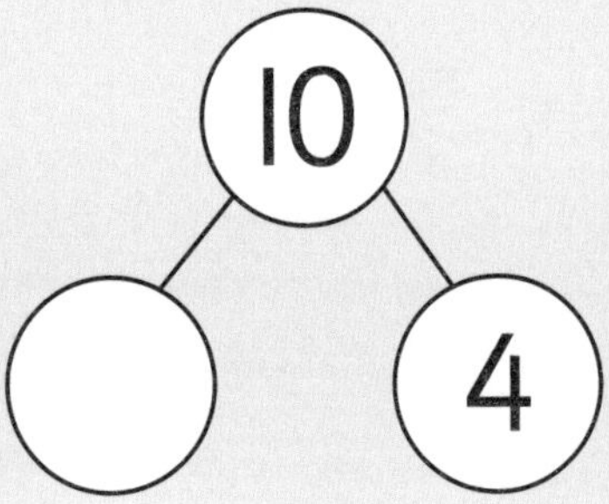

☐ + ☐ = 10

☐ + ☐ = 10

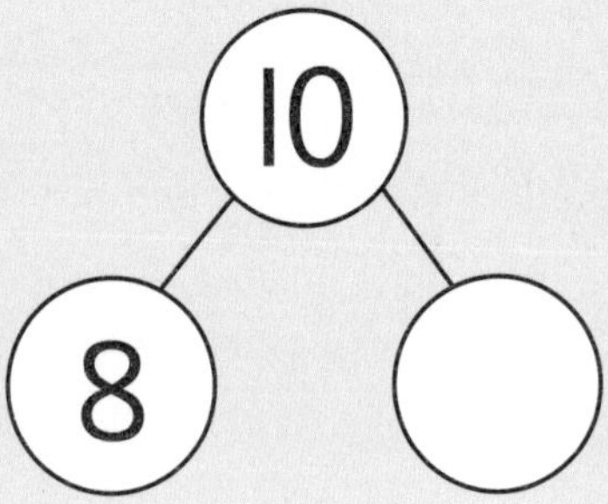

☐ + ☐ = 10

☐ + ☐ = 10

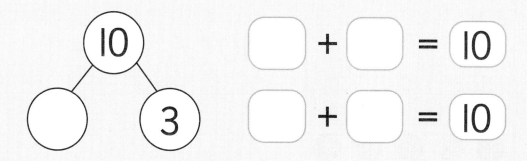

10

3

☐ + ☐ = 10

☐ + ☐ = 10

10

10

☐ + ☐ = 10

☐ + ☐ = 10

10

5

☐ + ☐ = 10

Teacher's notes

Complete each part-part-whole model: write the missing 'part'. Then write the addition number sentences to match.

Subtract from 10

10 − 9 = ☐

10 − 4 = ☐

10 − 6 = ☐

10 − 0 = ☐

10 − 3 = ☐

10 − 2 = ☐

$$10 - 7 = \boxed{}$$

$$10 - 1 = \boxed{}$$

$$10 - 8 = \boxed{}$$

$$10 - 5 = \boxed{}$$

$$10 - 10 = \boxed{}$$

Date: _____

Heaviest

 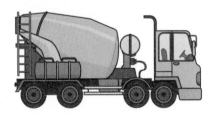

Balance scales

Date: _____

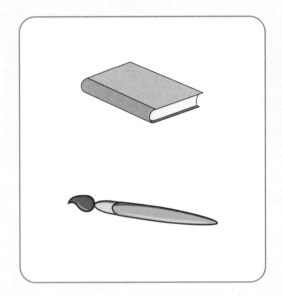

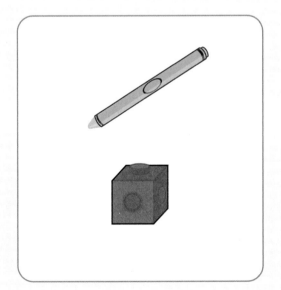

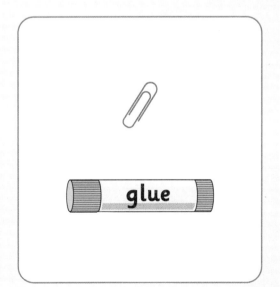

Teacher's notes

Place each pair of objects on a balance scale. Circle the lighter object.

21

Holds most

Date: _____

Teacher's notes

Circle the object in each set that holds the most.

Has most

Date: _____

Read and write numbers

Date: _____

1	2	3	4	5	6	7	8	9	10
11	12	13	14	15	16	17	18	19	20
21	22	23	24	25	26	27	28	29	30
31	32	33	34	35	36	37	38	39	40
41	42	43	44	45	46	47	48	49	50

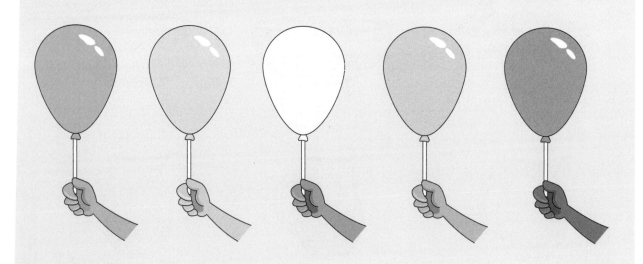

Teacher's notes

Top: Listen as an adult says a number. Find that number on the grid and draw a circle around it. They will say four other numbers for you to circle. Bottom: Listen as an adult says a number. Write that number on the first balloon. They will say four other numbers for you to write.

Date: _____

Ten and ones

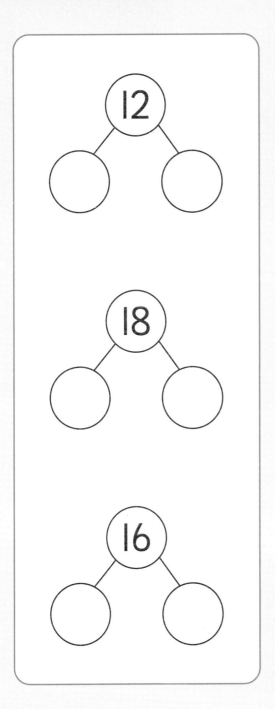

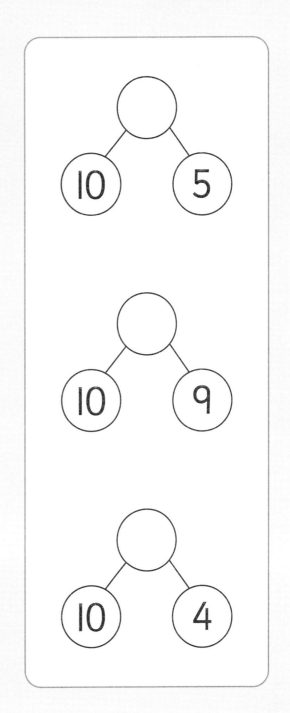

Teacher's notes

Orange set: Partition (decompose) each number into ten and ones. Blue set: Combine (compose) ten and ones.

25

Tens and ones

Date: _____

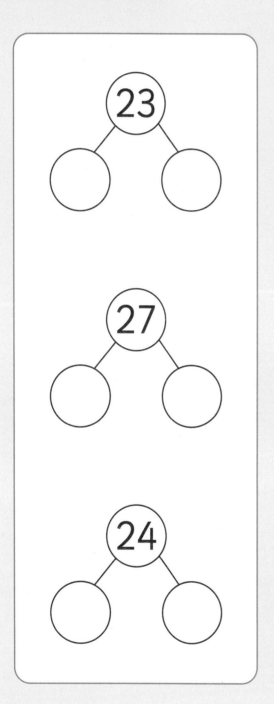

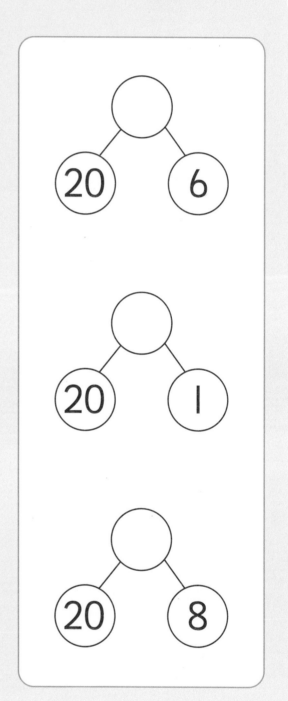

Teacher's notes

Orange set: Partition (decompose) each number into tens and ones. Blue set: Combine (compose) tens and ones.

Date: _____

Less

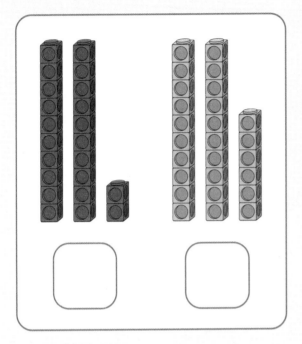

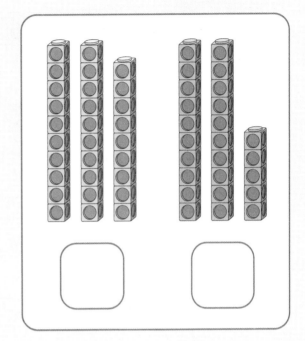

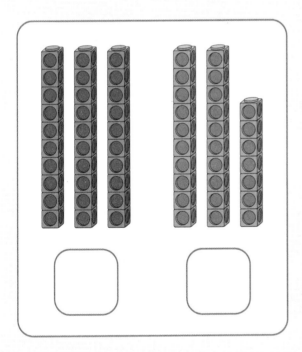

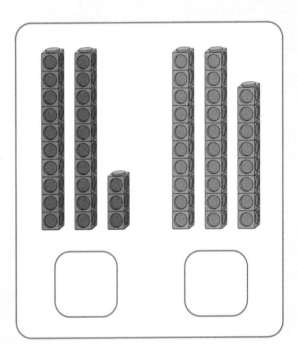

Teacher's notes

Count the cubes. Write the number in the box. For each pair, circle the number that is **less**.

27

Date: _____

Smallest and largest

0 1 2 3 4 5 6 7 8 9 10

11 12 13 14 15 16 17 18 19 20

21 22 23 24 25 26 27 28 29 30

14 5 17

20 10 16

9 13 21

30 28 22

25 29 19

23 27 26

Teacher's notes

For each strip of bunting, colour the **smallest** number **red**, and the **largest** number **blue**.

Smallest to largest

Date: _____

Teacher's notes

For each strip of bunting, an adult will show you three numbers. Write the numbers on the bunting from **smallest** to **largest**.

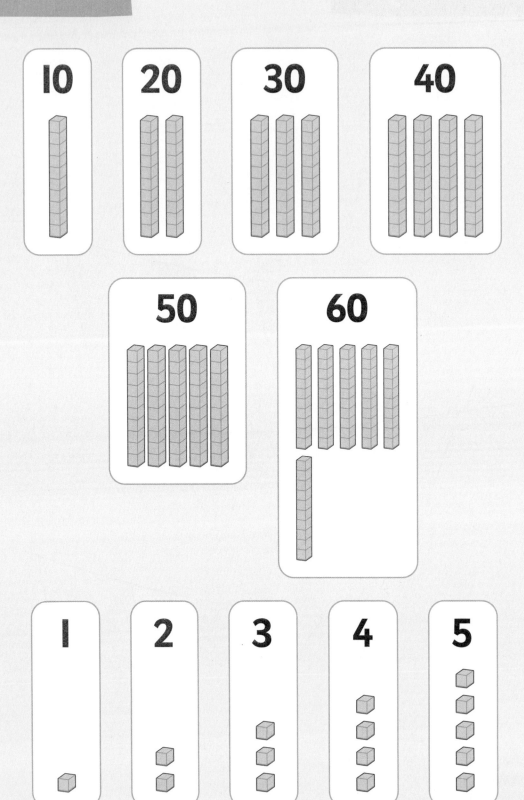

| 10 | 20 | 30 | 40 |

| 50 | 60 |

| 1 | 2 | 3 | 4 | 5 |

70

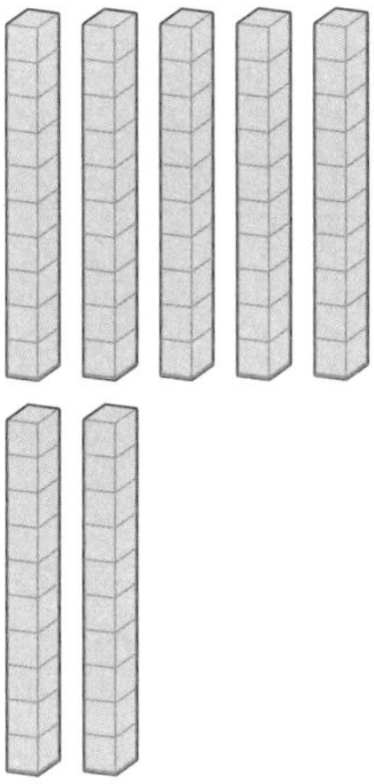

80

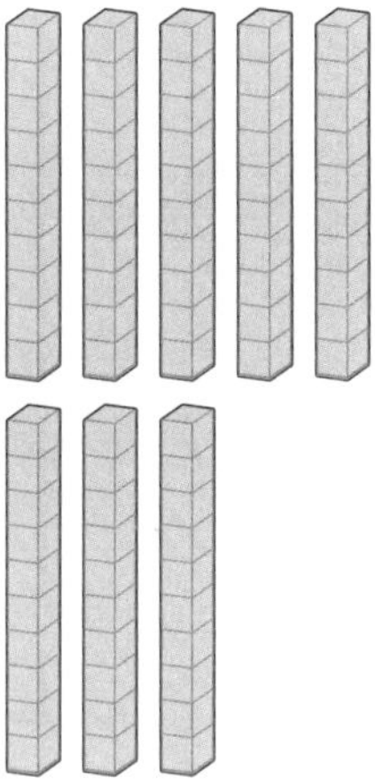

90

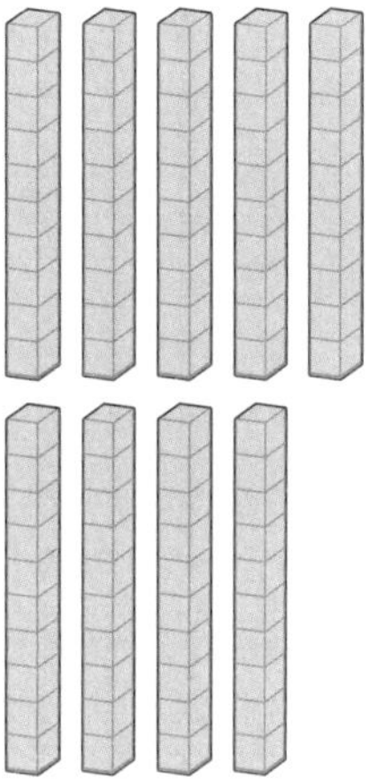

100

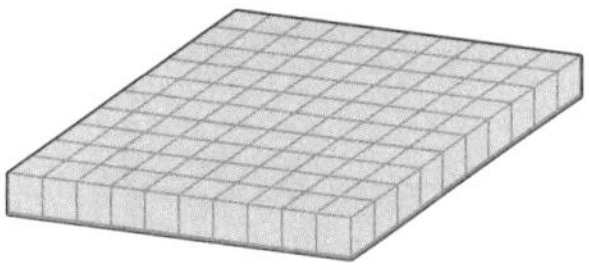

6

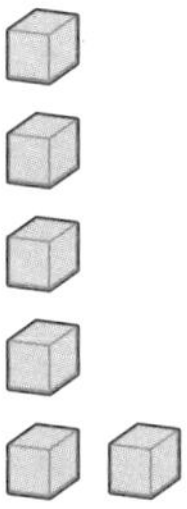

7

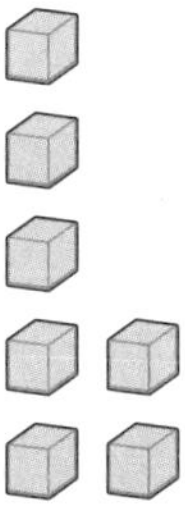

8

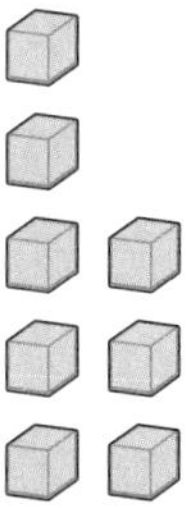

9

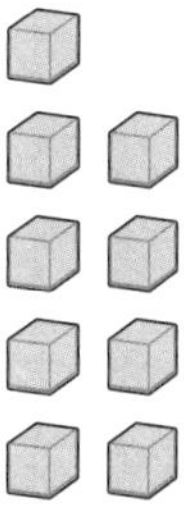

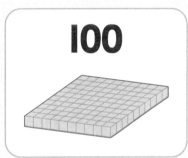

_____ has achieved these Maths Reception objectives:

Counting and understanding numbers
- Count on and back in ones, starting from any number
  from 0 to at least 30.   1   2   3
- Count objects, actions and sounds from 0 to at least 30.   1   2   3
- Recognise the pattern of the counting system for numbers from
  0 to at least 30.   1   2   3
- Compose (put parts together to make a whole) and decompose
  (break down a number into parts) numbers to at least 30.   1   2   3

Reading and writing numbers
- Read and write numbers from 0 to at least 30.   1   2   3

Comparing and ordering numbers
- Understand the relative size of quantities to compare numbers
  from 0 to at least 30.   1   2   3
- Understand the relative size of quantities to order numbers
  from 0 to at least 30.   1   2   3

Understanding addition and subtraction
- Automatically recall (without reference to rhymes, counting or
  other aids) number bonds up to 5 (including subtraction facts)
  and some number bonds to 10, including double facts.   1   2   3

Measurement
- Use everyday language to describe and compare weight, including
  heavy, heavier, heaviest, light, lighter, lightest, more and less. 1   2   3
- Use everyday language to describe and compare capacity
  and volume, including more, most, less, least, full, nearly full,
  empty and nearly empty.   1   2   3

1: Emerging   2: Expected   3: Exceeding

Signed by teacher:
Signed by parent: Date: